My name is Paul. I have two jobs. One job is teaching art classes. I do this job because I enjoy teaching.

I show the children good paintings. They tell me what they see.

We look closely at one part. Then they draw what they saw.

I help them with
their drawings. I teach them to
paint too. We mix colors.

The children paint because
it is fun.

My other job is even better. That is because I get to draw! I pack up my paints. I pause to grab my hat.

On nice days I go out at dawn. I go early because the light is soft.

I look for the best spot. Soon I start to draw.

One August day I saw a fawn. It did not see me at first. It crossed the lawn slowly. I took a snapshot of the fawn.

Later, I wanted to draw the fawn. I looked at the snapshot. That helped me draw the fawn.

I like this drawing a lot. Should I sell it? Should I keep it?

I think this drawing is just for **me.**

The End